Table of Contents

Foreword

> This instrument can teach, it can illuminate; yes, it can even inspire. But it can do so only to the extent that humans are determined to use it to those ends. Otherwise it is merely lights and wires in a box.
> —*Edward R. Murrow*

The oil crises of 1973–74 and 1978–79 shocked Americans economically and psychologically more than any non-military event since the Great Depression of the 1930's. To most Americans, the first crisis seemed to leap full grown from nowhere on October 17, 1973 when the Arab oil-producing states announced an embargo of petroleum shipments to the United States and reduced production in an attempt to gain Western support in the Arab-Israeli War. In the following months as prices quadrupled, the effects of this action were felt by the man-in-the-street from Maine to California. Although the embargo was lifted on March 19, 1974, television coverage of the effects continued into May 1974.

The second crisis, resulting from the Iranian Revolution, was felt in the United States in November 1978 as shortages began to develop. Iran, the world's fourth largest oil producer, ceased production by December 26, 1978. Saudi Arabia, the number one producer, exacerbated the Iranian cutoff by reducing its production by two million barrels a day in January 1978. In March 1979 OPEC raised the price of its oil by 40%; by year-end the price was doubled. In July 1979 Saudi Arabia increased its production by one million barrels a day and the crisis and television coverage wound down in August 1979.

How well did television teach and illuminate for its viewers the causes, effects and possible solutions to this major story which saw world oil prices jump eighteen-fold from less than $2 to about $35 a barrel in six years? How well did it outline the range of public policy issues and contribute to a better understanding of the public policy debate? How well did it serve the public's need to know?

This Media Institute study does not attempt to document the actual events that constituted the oil crises, nor to assess either the accuracy of the events or the relative merits of different explanations that appeared on television. Instead, we attempt to present a picture of what the three networks told the American public during 39½ hours of evening news stories that spanned the two crises.

In presenting this picture, it became clear that television may have contributed to the public confusion about causes and solutions—confusion reflected in public opinion polls taken during the 1970s. As Senator Charles Percy succinctly stated in 1974, "The American public doesn't know who or what to believe."

Needless to say, the television networks did not manufacture the frustration and confusion over oil shortages and price increases which gripped our nation. They reported, albeit with selectivity, a wide range of facts and opinions from government officials, elected representatives, consumers, the oil companies and many others whom they decided to interview. Television presented the news about the oil crises dramatically, wrestling with the continuing dilemma of being not only informative but entertaining. Large audiences that are not entertained are unlikely to listen, which serves no one's interest.

What was clearly lacking, however, were two essential elements that are critical to an understanding of any large economic event. First was historical perspective, which in the case of the oil crises would have included a perspective on international crude oil production, our domestic oil indus-

TV Coverage
of the Oil Crises:
How Well Was the Public Served?

Volume I
A Quaiitative Analysis

1973-74/1978-79

Edited by
Leonard J. Theberge

(**the media institute**)

ISBN: 0–937790–07–9

Library of Congress Catalog Card Number: 81–86030

try, and existing U. S. government policies. Without this history the relationship between these three dynamic elements was never clear. Secondly, television never adequately explained the law of supply and demand in regulating a dynamic economy. The most elemental concept in an introductory economics course has escaped many of our brightest and best informed citizens and it was undoubtedly a difficult subject to popularize. For a nation which relies so heavily on the role of prices in regulating and allocating its scarce resources, the lesson has proved almost impossible either to understand or to accept.

During the crises many figures in government and Congress assumed that the only acceptable way Americans would conserve energy was by government actions to reduce speed limits, turn down thermostats, and allocate scarce supplies. Few seemed to recognize that price has a profound influence on demand and supply. Television relied on independent experts to expose this point of view less than 2% of the time.

Without a historical or economic context within which to evaluate the morass of conflicting data and points of view, television taught us very little about the causes of and solutions to the energy crises.

There are many reasons why television network evening news is not the ideal medium for explaining complicated events. We can all sympathize with Walter Cronkite when he laments the difficulty in compressing ten minutes of information into one minute of air time. Yet does this explanation answer how the networks would handle the issue in ten minutes, in one hour or in thirty-nine hours?

If television news has failed to meet the challenge to "illuminate," it may be that we expect too much from it. Indeed, it is a marvelous technological medium that brightens the lives of many and imparts important if abbreviated information on events in the world. Television provides a healthy escape valve for society to vent its frustrations about events largely beyond its control. But does TV inform the public when, in its pursuit of balance, it presents a helter-skelter of viewpoints, some reasonable, others not, but very few that are disinterested and informed?

Our study on the oil crises is in three parts. Volume I is an analysis of the substance of over 1,400 oil crises stories; Volume II is a quantitative comparison of coverage among the networks; and Volume III is an analysis and commentary by economist Thomas W. Hazlett on how well network coverage mirrored the reality of the crises as they are understood today.

The Media Institute undertook this study as a sequel to its earlier study, *Television Evening News Covers Nuclear Energy: A Ten Year Perspective.* It is contemplated that additional studies on television coverage of various energy issues will be undertaken so that a comprehensive picture of energy coverage by television will be available for students, scholars and laymen interested in this essential area of media coverage.

None of the attributions contained in this report are intended to praise or blame individual reporters or networks. We respect the ideals of great broadcast leaders such as Edward R. Murrow and we are mindful of the human faults that prevent all institutions in our society from fulfilling their own ideals.

Leonard J. Theberge
President
The Media Institute

[W]hen the history books are written, however, it now seems likely they will describe the 1980's as a period of major adaptation, a time when oil and other cheap energy sources ceased to be regarded as social entitlements and as commodities somehow immune from the laws of economics.

—*Robert D. Hershey, Jr., in*
"Winning the War on Energy,"
The New York Times,
October 11, 1981.

Summary and Conclusions

Summary

The two oil crises of the last decade (1973–74 and 1978–79) were given extensive and prominent coverage on the evening news programs of the three television networks. Almost three-quarters of that coverage (74%) dealt with the causes of, solutions to, and effects of the crises. Based on a computer-aided content analysis of all oil crisis stories broadcast (1,462 stories totalling 39½ hours), researchers found the networks portrayed the oil crises in the following way:

- Non-market solutions to the oil crises such as conservation, rationing, regulation and price controls received three times the coverage that market solutions did—63% to 21%.
- Only 15% of all solutions discussed dealt with the major policy issue of regulation and price controls vs. deregulation and decontrol.
- Price was depicted as a problem rather than as a possible solution.
- The largest single source of information overall was the government—56%. By contrast, outside experts were the source of information only 2% of the time.
- The networks used government sources to discuss solutions 77% of the time. The oil industry was called upon to discuss solutions 9% of the time.

- When the networks addressed the subject of causes, they relied on government sources 53% of the time. Oil industry sources were called upon for 17% of the discussion.
- The networks identified government as a possible cause of the crises in only 18% of discussions of causes. By contrast, the networks identified OPEC and the oil industry as possible causes of the crises 72% of the time.
- Fully 25% of all causes discussed blamed the oil industry for perpetrating a hoax, profiteering, withholding supplies, or other devious actions.

Conclusions

The above findings raise very disturbing questions about the networks' performance in covering the foremost public policy issue of the 1970s. One conclusion stands out:

Network coverage failed to give the viewing public sufficient information to make an informed judgment about the oil crises.

This failure stemmed from two principal and interrelated shortcomings in the networks' coverage—failure to present adequately the economic dimension of the crises, and an overwhelming preference for government sources of information rather than outside experts. Let us consider these in turn:

The networks paid scant attention to the real debate, and in fact the central political and economic policy issue at stake: non-market solutions (rationing, regulation, price controls) vs. market solutions (deregulation, decontrol). The networks generally presented the solution to the oil crises as a choice among non-market measures: conservation vs. rationing vs. forms of regulation and price control.

The result was a distorted picture in which certain solutions became the focus of discussion, and in effect became the limits of the debate. Price mechanisms, which many economists and outside energy experts even then were discussing as the solution, were left out of that debate for the most part. The public, then, was led to believe that half the picture was in fact the whole picture.

The networks demonstrated an over-reliance on government sources for information about the causes of and solutions to the oil crises.

The networks turned to the government for information far more often than to any other source. This weakness cannot be dismissed by saying the networks were merely reporting the news which happened to originate with the government. One need not have a political or economic point of view to recognize the imbalance and asymmetry in the use of government and non-government sources, and the imbalance in the presentation of causes and solutions. (Government was the most frequent source of information, but was portrayed infrequently as the cause of the crises.)

This is not a new phenomenon, however. In a previous study on television coverage of the inflation issue*, it was found that government was the source of 77% of inflation stories, and that government was exonerated as a cause of inflation in 77% of stories broadcast.

*   *   *

The findings of this study place a burden on the networks to explain why the government's role in causing and prolonging the oil crises received such little coverage, and why deregulation and price decontrol were given such little attention as a solution.

The burden is especially heavy regarding this latter point. Since the end of the last crisis, there has been a notable improvement in the domestic energy picture. While the causes of and solutions to the oil crises are still a matter of considerable public debate, economists have long held (as more journalists lately recognized) that the principal cause was the oil industry's regulated environment, and that the

---

* Bethell, Tom, *Television Evening News Covers Inflation: 1978–79*, The Media Institute, Washington, D.C. 1980.

solution lay in deregulating the industry, facilitating greater domestic oil production, and in general relying on the market as the best distributor of scarce commodities.**

** For more on this topic, for example:

Steele, Henry, 1969. Statement before U.S. Senate, Judiciary Committee. *Hearings on Government Intervention in the Market Mechanism: The Petroleum Industry.* Washington, DC: Government Printing Office, pp. 208–222.

McDonald, Stephen L., 1971. *Petroleum Conservation in the United States: An Economic Analysis.* Baltimore: Johns Hopkins University Press.

Adelman, Morris A., 1972. *The World Petroleum Market.* Baltimore: Johns Hopkins University Press.

Mitchell, Edward J., 1974. *U.S. Energy Policy: A Primer.* Washington, DC: American Enterprise Institute.

Institute for Contemporary Studies, 1975. *No Time to Confuse,* San Francisco, CA.

Friedman, Milton, 1975a. Two Economic Fallacies, *Newsweek* 12 May:83.

Friedman, Milton, 1975b. Subsidizing OPEC Oil, *Newsweek* 23 June:75.

RAND Corporation, 1977. *Petroleum Regulation: The False Dilemma of Decontrol.* Santa Monica, CA: RAND Corporation, January.

For discussion of the analogous subject of natural gas see, for example:

MacAvoy, Paul W., 1971. The Regulation-Induced Shortage of Natural Gas, *Journal of Law and Economics* April:167–199.

Erickson, Edward W., 1971. Supply Response in a Regulated Industry: The Case of Natural Gas, *Bell Journal of Economics and Management Science* Spring:94–121.

More generally, see:

Hotelling, Harold, 1931. The Economics of Exhaustible Resources, *Journal of Political Economy* April: 137–175.

Herfindahl, Orrus C., and Allen V. Kneese, 1974. *Economic Theory of Natural Resources,* Columbus, OH: Merrill Publishing Co.

For general discussion of how price controls result in shortages see, for example:

Samuelson, Paul A., *Economics,* McGraw-Hill (New York, New York: 1973, 9th edition) pp. 392–395.

Hirschleifer, Jack, *Price Theory and Applications,* Prentice-Hall (New Jersey: 1976) p. 265.

Alchian, Armen A. and William R. Allen, *University Economics,* Wadsworth Publishing (Belmont, CA: 1972, 3rd edition) pp. 100–101.

Cheung, Steven N. S., "A Theory of Price Control," *Journal of Law and Economics,* Vol. 17, No. 1, April 1974.

For recent articles on oil economics see, for example:

Tucker, William, "The Energy Crisis is Over," *Harper's,* November 1981.

Hershey, Robert D., Jr., "Winning the War on Energy," *The New York Times,* October 11, 1981.

Introduction

The oil crises of the 1970s had an enormous impact on American economic priorities, policies and lifestyles. Until the gas lines formed, few Americans knew or cared about energy. It had always been there, and it had always been cheap. The crises caught the public, and most of the news media as well, by surprise. What were these crises, what caused them and what solutions were available? To explain the oil crises to the public, television first had to reach its own understanding of the crises. How well did the networks meet that challenge?

The answer is important, because television is the main source of news for nearly two-thirds of the American public.[1] Moreover, Americans regard television as by far the most believable of the news media.[2] For that reason alone, television's portrayal of the oil crises must be considered influential. What was the nature of the oil crises, then, according to television?

The focus of this study is not the oil crises themselves, but rather how these crises were covered on the network evening news programs.

---

[1] The figures ranged from 60% in 1971 to 67% in 1978. *Evolving Public Attitudes Toward Television and Other Mass Media 1959–1980*, Roper Organization Inc., p. 3.

[2] *Ibid.*, p. 4. Television is favored by a two-to-one margin over newspapers in this respect.

In the interest of thoroughness, researchers analyzed every pertinent story [See Appendix for story selection criteria] broadcast during the periods of October 1973–May 1974 (an eight-month period) and November 1978–August 1979 (a ten-month period). Over 39 hours of news—more than 1,400 stories—were examined. Using a quantitative research technique called content analysis, the Institute collected and analyzed, using computers, a massive amount of data which yields a number of significant insights into network news coverage of the oil crises.

The effects of this coverage on the perceptions of the television audience is beyond the scope of this study. While television may or may not have a major influence on public opinion, it plays an important role in setting the agenda for national debate. As Bernard Cohen observed: "The mass media may not be successful in telling us what to think, but they are stunningly successful in telling us what to think about."[3]

This study examines, among other things, the agenda set by television in its coverage of the oil crises. While it is important to determine what television said about particular subjects, of greater importance is the simple matter of which issues television chose to cover at all—because the media function as an information gatekeeper, selecting from the vast assortment of ideas and opinions those few that will enjoy wide exposure.

In addition, the content of television coverage is significantly influenced by the sources of information used. Therefore, this study also analyzes the sources relied on by the networks and how those sources were used.

This analysis is the first in a three-volume study The Media Institute is publishing on network television evening news coverage of the oil crises. The second volume will analyze the quantitative differences among the networks in regard to their coverage—how ABC, CBS and NBC differed in the way each chose to cover the oil crises, and how the coverage changed from the first crisis to the second. The

---

[3] Bernard Cohen, *The Press and Foreign Policy* (Princeton, N.J.: Princeton University Press, 1963).

will offer an economic analysis of the accuracy, relevance and balance of television's portrayal of the oil crises.

These three volumes share a common, two-fold purpose: to contribute to an understanding of how television reports on a major policy issue; and to identify specific areas in which the quality of television news might be improved.

I. General Dimensions of Coverage

It would be helpful to begin by considering story subjects and sources which are the foundation for later analyses. For this study, 1,462 stories were analyzed—a total of 39½ hours of coverage. As Graph 1 shows, 24 hours of this coverage occurred during the first crisis, some 61% of the total. Coverage was also more concentrated in the first crisis [See Graph 2] and at times accounted for more than half the news in a single broadcast.

Subjects

Because a single story might cover several different subjects or present several opposing views on the same subject, this study was designed to accomodate such complexity and variety. (For the purposes of this study, a "subject" was defined as an issue or point of view raised in the oil crises stories, such as oil-crisis-induced unemployment or lowered highway speeds.) Whenever a subject was raised, or when opposing viewpoints on the same subject were presented (such as a story in which one Congressman advocated rationing and a second opposed it), researchers noted each subject and its source. Researchers found about four subjects in the average story, with each subject accounting for an average of 25 seconds of news. Of course, some stories contained only one subject, while very long and complex stories contained as many as 20 subjects. In total, the 1,462

Graph 1

Total Coverage, in Stories, in Hours, and in Subjects

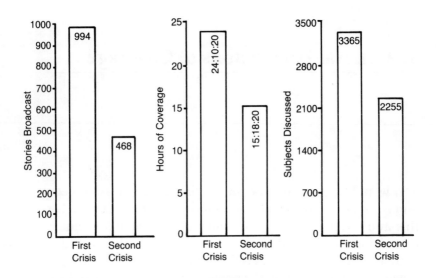

Graph 2

Coverage of Oil Crises Per Month

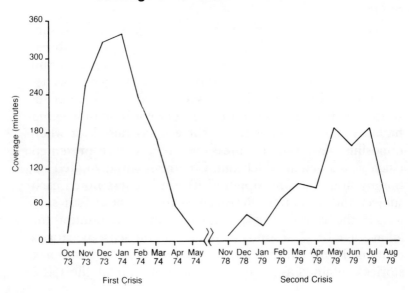

stories contained 5,620 separate instances in which subjects were raised.

Three-quarters of these subjects could be classified as concerned with solutions to, causes of, or effects of the crises. The remaining quarter was not discussed in this context, and dealt instead with the technical data of supply levels, actual prices or price increases, profit reports of the oil companies, production levels in the United States and abroad, and legal matters concerning the oil industry. [See Chart 1]

Sources

For each subject raised, researchers noted the "source" of the information, that is, the person stating the information or to whom the information was attributed. Such sources were then aggregated according to their affiliation (*e.g.*, government, oil industry, non-oil industry, OPEC). Frequently, however, it was not possible to determine the source of a particular subject because the correspondent did not identify his source, was editorializing, or was stating a commonly held view. For example, consider a report from Vienna concerning the outcome of an OPEC meeting. When NBC's John Dancy noted that higher oil prices would have an effect on inflation, it is unclear whether the conclusion was his own or whether he relied on another source:

> So when the Arab oil begins to flow again, it will be more expensive. That means higher prices for petroleum-based products and that in turn means more inflation, which the oil-producing nations say they don't want, but which they are in fact helping to create. John Dancy, NBC News in Vienna.[4]

Whenever a subject had an unattributed source, and the source could not be determined with a high degree of confidence by the researcher, the source was coded as unidentifiable. About 18% of all subjects came from sources that could not be identified with a high degree of confidence.

The remaining 82% of subjects had identifiable sources.

---

[4] March 17, 1974.

Chart 1

Subjects Discussed
[*100% = 5620 subjects*]

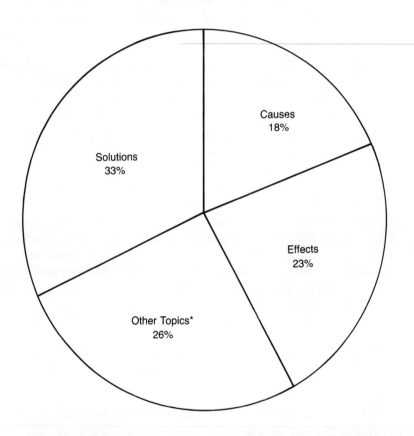

*includes discussion of supply levels and shortage, prices, oil industry profits, etc., when these subjects were not discussed as solutions, causes, or effects

Chart 2

Sources* of Information
[100% = All Identifiable Sources = 4585 Sources]

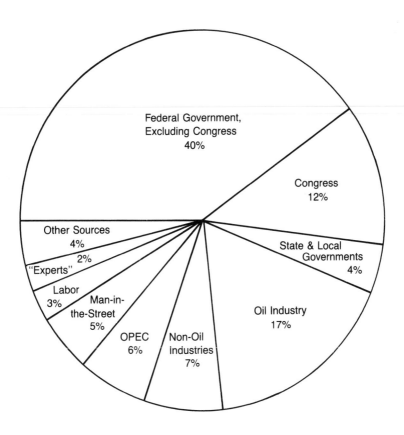

Federal Government, Excluding Congress 40%

Congress 12%

State & Local Governments 4%

Oil Industry 17%

Non-Oil Industries 7%

OPEC 6%

Man-in-the-Street 5%

Labor 3%

"Experts" 2%

Other Sources 4%

*excluding unidentifiable (unattributed) sources

Graph 3

**All Sources and the
Subjects They Discussed**

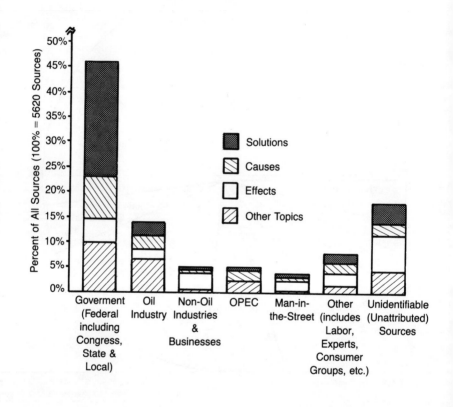

In oil crises coverage, government sources (*i.e.*, Federal, state and local) accounted for 56% of all identifiable sources. The oil industry, at 17%, was the next largest source of information. A quarter of these oil industry sources were service station operators.[5] Non-oil industries and businesses, encompassing a cross-section of American business, accounted for another 7%. The Organization of Petroleum Exporting Countries (OPEC) comprised another 6% of all identifiable sources. The "man-in-the-street" (*i.e.*, the individual who was sought out for his personal view) accounted for another 5%. Representatives of labor, such as autoworkers, teamsters and miners, constituted 3% of all identifiable sources. "Experts" (*e.g.*, financial analysts, economists, energy experts) totaled 2% of identifiable sources. The remaining identifiable sources did not fit into any of these categories and were aggregated under the category of "other" sources. Included in this group were the American Automobile Association, consumer groups, foreign leaders, foreign countries (particularly Canada and Mexico), the Gallup Organization, Associated Press, and academic institutions. [See Chart 2]

The above breakdown gives no indication of what subjects these sources discussed. A measure of this is provided in Graph 3. It is evident that some sources were used extensively in one type of discussion, yet excluded from others. Researchers found that 33% of all subjects were concerned with solutions and 23% with effects. However, solutions accounted for 50% of all subjects from federal, state and local government sources, and effects only 10%. By contrast, only 10% of the subjects from non-oil industries dealt with solutions, while 73% dealt with effects. It is also interesting to note that the greatest proportion of the subjects with an unidentifiable source dealt with the effects of the crises. Such disparities will be considered in greater detail in subsequent chapters.

---

[5] Oil company executives and service station operators do not necessarily share the same views. But because operators are an important element of the industry's retail segment, they were included in this category.

II. Solutions to the Crises as Portrayed by the Networks

The most striking feature of the network's coverage of the oil crises was the mix of solutions presented. Fully 50% of the discussion focused on conservation (excluding price-induced conservation) and rationing. In contrast with the expansive attention given to non-market solutions, the coverage afforded market solutions seems somewhat meager.

This is the case, despite the fact that the price/consumption relationship is almost universally accepted. This elementary economic fact of life states that a society will obtain the most utility from its resources by allowing consumers to "bid" against each other for the use of each commodity. The bid takes the form of the price consumers are willing to pay, and through the price mechanism commodities are allocated throughout an economy in proportions reflecting greatest utility. Both rationing and conservation, on the other hand, are systems that ignore the market function of prices.

Solutions that acknowledged the role of the market, such as the price-mechanism, decontrol and deregulation, and domestic production, together accounted for only 21% of subjects discussed.[6]

Moreover, while one of these market solutions, deregu-

---

[6] Oil was the one commodity that continued to have a controlled price after President Nixon's 1971 price freeze was lifted.

lation and decontrol, was discussed 4% of the time, the opposite non-market solution of regulation and price controls accounted for 11% of the discussion of solutions— nearly three times as much. [See Chart 3]

Many economists contend that there would have been no oil shortage and no gas lines had the price of oil been permitted to reflect its relative scarcity rather than being artificially kept low. Yet *the entire debate regarding regulation of the oil industry versus deregulation, and price controls versus decontrol, constituted only 15% of the solutions discussed.*

The agenda set by the networks for the discussion of solutions to the oil crises largely neglected the role of prices. Instead, the options presented tended to be one of the several offered in the following report from ABC's Roger Peterson:

> Economists who study the oil situation are divided on whether they think *government action* will become necessary. But there is general agreement on one point: If Americans would start using their common sense and *stop wasting energy,* we wouldn't be faced with *mandatory conservation,* much less *rationing.* [7]

As ABC's Charles Gibson noted in one report, "Gas lines, of course, are by themselves a form of rationing because not everyone can afford the time to endure them"[8] With that in mind, the options recounted that same evening in an NBC story by John Chancellor seem narrow indeed: "[Alfred] Kahn said the American people will prefer rationing to waiting continually in long gasoline lines".[9]

A similarly narrow choice, this time between voluntary and mandatory rationing, was offered by CBS' Richard Wagner in a story about the "Oregon Plan," a voluntary odd/even day rationing system: "... the Oregon experiment will doubtless be watched carefully by energy officials around the country to see if a strictly voluntary plan can work and thus make compulsory rationing of fuel a step the nation won't have to take."[10]

---

[7] February 27, 1979 (emphases added).
[8] June 28, 1979.
[9] June 28, 1979.
[10] January 14, 1974.

Chart 3

Solutions Discussed
[*100% = 1880 subjects*]

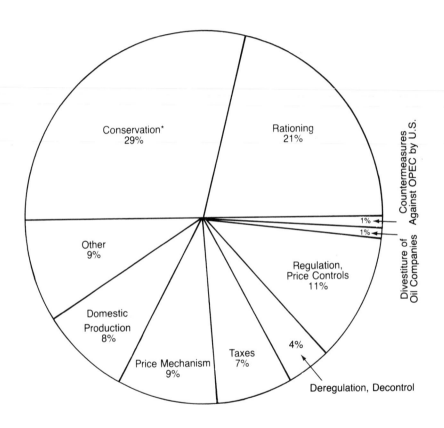

Conservation*
29%

Rationing
21%

Countermeasures
Against OPEC by U.S.

Other
9%

1% ←

1% ←

Divestiture of
Oil Companies

Regulation,
Price Controls
11%

Domestic
Production
8%

Price Mechanism
9%

Taxes
7%

4%

Deregulation, Decontrol

*excluding price-induced conservation

Not all discussion of rationing favored it as a solution. One-fifth of the discussion dealt with the question of whether or not there would in fact be rationing. Another fifth specifically rejected rationing as a solution. For instance, William Simon, then head of the Federal Energy Administraiton, was shown on NBC saying, "I consider rationing absolutely full of inequities ... there are those that say that gasoline rationing is inevitable and I will not accept that."[11]

Conservation (not including price-induced conservation) received more coverage than any other solution—a full 29% of all solutions discussed. Discussion of conservation as a solution included both its mandatory forms, such as lowered highway speeds, and voluntary ones, such as car pooling. In lieu of the inducement of price to spur conservation, appeals were made to patriotism and to a sense of shame. President Nixon described his program as "one in which everyone will sacrifice something, but ... no one will be required to suffer."[12] Following the President's request that gas stations remain closed on Sundays, ABC's Anne Kaestner asked a skier in California, "Do you feel guilty buying gas on Sundays?"[13]

About one-third of the time that conservation was discussed as a solution, it was discussed in general terms. [See Chart 4] Federal Energy Administration chief John Love was shown on ABC saying, "I believe that the voluntary and mandatory conservation practices are going to have to be a large part of this because the alternative, in my mind, is so difficult and almost unacceptable."[14]

Another 17% of the discussion of conservation dealt with the effectiveness or success of conservation measures. For instance, Walter Cronkite began one broadcast with this:

> The Nixon Administration today announced more steps to conserve fuel and electricity, and it said that voluntary cooperation by citizens already has shown a significant

---

[11] December 5, 1973.
[12] CBS, December 13, 1973.
[13] December 17, 1973.
[14] December 8, 1973.

Chart 4

Conservation* Discussed as a Solution

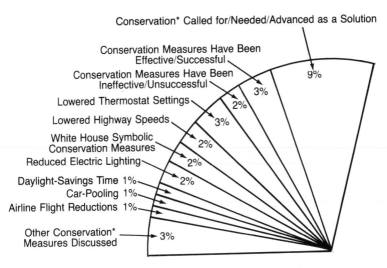

Conservation* Called for/Needed/Advanced as a Solution — 9%

Conservation Measures Have Been Effective/Successful — 3%

Conservation Measures Have Been Ineffective/Unsuccessful — 2%

Lowered Thermostat Settings — 3%

Lowered Highway Speeds — 2%

White House Symbolic Conservation Measures — 2%

Reduced Electric Lighting — 2%

Daylight-Savings Time 1%

Car-Pooling 1%

Airline Flight Reductions 1%

Other Conservation* Measures Discussed — 3%

*excluding price-induced conservation

Rationing Discussed as a Solution

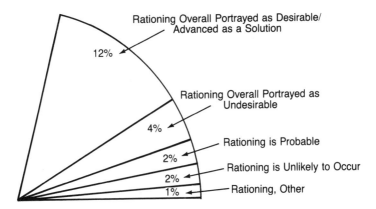

Rationing Overall Portrayed as Desirable/Advanced as a Solution — 12%

Rationing Overall Portrayed as Undesirable — 4%

Rationing is Probable — 2%

Rationing is Unlikely to Occur — 2%

Rationing, Other — 1%

cutback in gasoline use. If the programs work, the energy chief said, "I believe we will not need rationing."[15]

When specific forms of conservation were discussed, the solution of adjusting the thermostat to less comfortable temperatures received the most attention, particularly during the second crisis. The reduction of maximum posted highway speeds was another big story during the first crisis.

The White House received considerable coverage for symbolic acts of conservation in both crises, such as having only one light on the national Christmas tree and reducing the speeds of White House planes, helicopters and cars. President Nixon generated some controversy by taking a commercial flight to California, leaving the press to scramble for themselves. In another "fuel-saving measure"[16] the President cancelled the regular chartered plane for newsmen who wanted to cover his post-Christmas vacation in Florida.

Reduction in the use of electric lighting was another conservation measure discussed at some length. In particular, businesses were encouraged to reduce or forego their use of outdoor lights and neon signs. Another conservation measure which received substantial coverage in the first crisis was the enactment of America's first peacetime daylight-savings time law.[17] A controversy followed as to whether this actually saved any energy or resulted instead in harm to children on their way to school in the dark.

Fuel conservation was also discussed in terms of car pools and reductions in the number of airline flights. About 10% of the discussion of conservation as a solution fit into one of the above categories. Quite a vast array of conservation measures were considered in television's coverage. Among them, the production of more fuel-efficient cars was mentioned a number of times, as was the possibility of changing the semesters of schools and universities so that they could remain closed during the coldest winter months. Also discussed were other potential solutions, such as holding fewer

---

[15] December 13, 1973.
[16] According to Harry Reasoner, ABC, December 18, 1973.
[17] December 14, 1973.

sporting events, using only cold water to wash clothes and closing breweries.

Solutions involving regulation of the oil industry and controls on oil prices accounted for 15% of all solutions discussed—about half the share allotted to conservation as a solution. In the first crisis, almost all discussion of this subject was in support of the government's regulation of the industry. Senator Henry Jackson, for example, appeared on CBS saying: "What I am shooting for here is to get a leash, first of all, on the [oil] companies, so that there can be some sensible regulation...."[18]

NBC reported that "Senator Henry Jackson said today he's thinking about introducing a bill to require all [oil] firms which operate overseas to have a U.S. government official on their board of directors."[19]

Likewise, price controls were discussed without any indication that controls might be harmful or that they were anomalous in the economy as a whole. Thus Walter Cronkite reported that eight congressmen "urged a three-month freeze on oil prices,"[20] the governor of Massachusetts "called for public regulation, or if necessary, public ownership of the nation's oil companies,"[21] and the head of the Federal Energy Office urged the oil industry to cooperate with the government's fuel allocation program.[22] If such positions were controversial, the report gave no indication of it.

In the second crisis, coverage of regulation was somewhat different. First, the debate on regulation vs. deregulation was given more attention. Second, this coverage occupied a greater share of the total solutions discussed, increasing from 12% of all solutions in the first crisis to 21% in the second. However, the entire debate still received only one-half the coverage given to either conservation or rationing. It is significant that coverage of deregulation and decontrol received increased coverage only when the Administration endorsed these measures. The process of decontrolling oil

---

[18] January 11, 1974.
[19] March 17, 1974.
[20] January 16, 1974.
[21] *Ibid.*
[22] *Ibid.*

prices was begun by President Carter in June of 1979. As CBS's Eric Engberg explained: "The Administration argues that higher domestic prices will encourage consumers to use less and producers to find more."[23] On NBC, Charles DiBona of the American Petroleum Institute said, "The existing system of Federal regulation is causing all of the energy problem. If you eliminated that, you would eliminate the problem."[24]

However, opposition to decontrol was given equal time. On CBS, Ellen Berman of the Consumer Energy Council asserted that "decontrol will do virtually nothing to increase domestic production or to reduce consumption . . . it will not work."[25] And all three networks presented Kathleen O'Reilly of the Consumer Federation of America calling President Carter's decision to decontrol "the most in-flationary and anti-consumer White House decision in this century."[26]

Not long ago, *The New York Times* noted that "rising [oil] prices, it is becoming increasingly evident, are doing a job that exhortation, production controls, appeals to patriotism and plain gloom-mongering failed to do in curbing con-sumption."[27] If the price mechanism has ultimately proved to be the solution to the oil crises, the extent to which tele-vision set that option before its viewers becomes particularly interesting. Nine percent of all topics discussed as solutions concerned the price mechanism—under half the coverage given rationing, and less than a third of conservation's coverage.

One way in which the price mechanism was covered was in the context of price-induced conservation. For instance, NBC's Irving R. Levine reported that "[William] Simon wants to make rationing unnecessary by cutting down the demand for gasoline to match the available supplies and

---

[23] June 1, 1979.
[24] December 18, 1978.
[25] June 1, 1979.
[26] June 1, 1979.
[27] "Winning the War on Energy," *The New York Times*, Robert D. Hershey, Jr., October 11, 1981, Sec. 3, pp. 1–17.

Simon's energy experts say that to accomplish this, the price will have to go up to 80 cents a gallon...."[28]

During the second crisis, Levine reported that Alfred Kahn, head of the President's Council on Wage and Price Stability, told the Joint Economic Committee that "fuel prices must continue to go up in order to reduce consumption."[29]

Other coverage of the price mechanism evinced a certain bewilderment. NBC's Richard Hunt devoted four minutes to an investigation of whether Arab oil was slipping through the embargo to this country. He began by identifying a tanker that had been loaded in Turkey with gasoline. "There's no way of knowing whether it was made from Turkish or Arab oil,"[30] he noted. A second tanker was loaded with gasoline in Rumania. "Again, there's no way of knowing which kind of gasoline Rumania is selling to the United States."[31] Having established what might be in the tankers, Hunt followed the course of the gasoline to this country, where he discovered a gas station that might ultimately receive the gas. Hunt concluded that:

> The gasoline in this pump no doubt travelled a long way and passed through many hands to get here. Apparently every transaction was legal, but each step added something to the price. It seems to show that more gas is available, if people are willing to pay.[32]

The price-sensitivity of supply was also noted in a report by NBC's Jim Scott, who interviewed the operator of a Texaco station in Pittsburgh. "I think there's a lot of gas out there, but I think they just want to get the price up. I think once the prices are up, there will be all the gas we can handle," the operator said.[33] The viewer was left to wonder whether this was a solution or some kind of devious ploy on the part of the oil companies.

---

[28] December 5, 1973.
[29] December 6, 1978.
[30] March 13, 1974.
[31] *Ibid.*
[32] *Ibid.*
[33] March 6, 1979.

Another solution which received coverage was that of domestic production of oil. About 8% of the subjects discussed as a solution dealt with whether domestic production could alleviate the country's dependence on foreign oil. One month after the 1973 Arab oil embargo began, Congress quickly passed the Alaskan pipeline bill, which President Nixon then signed. While ABC's Roger Peterson noted that "passage of the pipeline bill won't help alleviate our current energy shortage,"[34] it would, according to President Nixon, "help solve the nation's long-range energy problems."[35]

Much of the coverage of domestic production in the first crisis focused on the possible tradeoff between environmental goals and increased domestic exploration, drilling and production. By the second crisis, the focus had turned to whether, in the wake of decontrol, domestic production could increase enough to reduce our dependence on foreign sources of oil. As Deputy Secretary of Energy John O'Leary told Irving R. Levine, "We don't have any more Alaskas to come in and help us out."[36]

By and large, the remaining topics discussed as solutions dealt in some fashion with government action. Discussion of taxes, to be placed on either gasoline or on oil industry profits, constituted 7% of all solutions discussed. The primary tax considered was the windfall profits tax, which accounted for 3% of all solutions in the first crisis and 12% in the second. Over 70% of the time, the windfall profits tax was portrayed as desirable. While CBS presented Rep. Barber Conable saying the tax bill "will kick the oil companies in the face,"[37] Rep. Al Ullman followed, asserting that "This is a good, tough, fair bill. It won't cost the taxpayer anything but it will take away the windfall...."[38]

The solutions covered in the first crisis included discussions by Secretary of State Kissinger, Defense Secretary Schlesinger and Vice President Ford, among others, of the possibility of using either economic or military reprisals

---

[34] November 13, 1973.
[35] ABC, November 12, 1973.
[36] NBC, June 1, 1979.
[37] June 1, 1979.
[38] *Ibid.*

against the OPEC members participating in the oil embargo. In both crises, some coverage was given to "anti-monopoly"[39] legislation to "dismantle the structure of the big oil firms."[40] Walter Cronkite went so far as to move from behind the anchor desk so that he could maneuver large building blocks to illustrate the meaning of "vertical integration."[41]

Some solutions that were covered do not fit into any of the categories previously discussed. Much of this was in the form of general discussion of an energy program or policy. Sometimes this program or policy was embodied in a bill, such as President Nixon's Emergency Energy Bill, or President Carter's Emergency Conservation Bill. Both bills combined a wide variety of solutions, including rationing and alternate fuels,[42] and stories reporting the course of such bills through the legislative process often did not even mention the bills' individual components.

Considerable coverage was also devoted to President Nixon's efforts at organizing international conferences of oil-importing nations. Much discussion centered on the possibility that the crisis could be alleviated if the oil consuming nations presented a unified front to OPEC. Both Nixon and Kissinger appeared on the news numerous times, exhorting our allies to keep up a common front. In January 1974, France cut a 20-year oil deal with Saudi Arabia in exchange for jet fighters and heavy arms, leading to the eventual abandonment of this solution, at least from television's viewpoint. Other proposed solutions included the relaxation or suspension of clean air standards, tapping the Defense Department's oil reserves, and expanding the services of the nation's railroads. Still another unclassifiable solution to receive coverage was an amendment attached in 1973 by the House of Representatives to the Emergency Energy Bill.

---

[39] NBC, March 27, 1974.
[40] November 29, 1973.
[41] *Ibid.*
[42] As the subject of this study was the oil crises rather than an energy crisis more generally, the myriad issues raised by alternate fuels were not included in this study. Television's coverage of alternate sources of energy would be a fascinating and worthwhile area for future research. See *Television Evening News Covers Nuclear Energy: A Ten Year Perspective,* The Media Institute (1979).

That proposal would have prohibited the use of gasoline under the allocation system for busing students beyond neighborhood schools.

* * *

Television consistently limited its viewers to choosing among non-price solutions. The availability and price of oil were depicted as twin problems (*e.g.*, "All across the country, oil is not only expensive, it is in short supply,"[43]) rather than as a problem and its solution. If energy economists continue to debate the ultimate solution (or combination of solutions) to the oil crises, the same cannot be said for television. With near single-mindedness, television covered a narrow set of solutions that featured the government and ignored the price mechanism. How did it come to set the agenda in this way?

While this study could not determine why television chose the information it did, researchers did note the source of each subject to appear on the news. When solutions were discussed, 77% of the subjects came from government sources.

Correspondent Bill Moyers recently observed that, "Most of the news on television is, unfortunately, whatever the government says is news."[44] In oil crises coverage, a solution merited consideration rarely unless it was advanced by the government. And the government, according to network coverage, was going to solve the crises by: (1) organizing consumer demand; (2) getting control of the oil industry; and (3) standing up to or co-opting OPEC. It is ironic that many economists and other experts contend the government solved the oil crises only by withdrawing from, and deferring to, the market.

The government's dominance of the discussion of solutions actually increased in the second crisis, rising from 73% of all identifiable sources to 82%. This increase was due to the nearly two-fold increase in the role of Congress as a source—from 14% of all sources in the first crisis to 26% in

---

[43] ABC, February 28, 1979.
[44] "The New Look at CBS News," *The Washington Post,* December 23, 1981, p. C-3.

Chart 5

Sources* for Discussion of
Solutions to Crises
[*100% = 1645 Identifiable Sources—see footnote*]

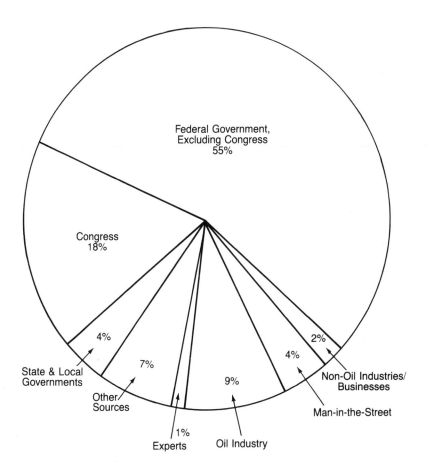

Federal Government,
Excluding Congress
55%

Congress
18%

4%

State & Local
Governments

7%

Other
Sources

1%

Experts

9%

Oil Industry

4%

Man-in-the-Street

2%

Non-Oil Industries/
Businesses

*unattributed sources excluded (235 solutions, or 12% of all solutions discussed,
were provided by unidentifiable sources)

the second. The oil industry was largely cut out of the discussion of solutions in the second crisis. It was the source of 11% of all solutions in the first crisis, and only 6% in the second.

Chart 5 gives a breakdown of the sources for discussion of solutions. The "other" category includes such sources as conservation groups, OPEC, labor and environmental groups.

III. Causes of the Crises as Portrayed by the Networks

What agenda did television set for the discussion of the cause(s) of the crises? Overall, actions of OPEC and the oil industry accounted for over 70% of all causes discussed. [See Chart 6] In the first crisis, OPEC and the Arab oil embargo were the subject of 43% of all discussion of causes, with the oil industry next at 32%. In the second crisis, when oil production in Iran was halted by revolution, discussion of the oil industry as a cause accounted for 41% of all causes discussed, while OPEC and the Iranian cutbacks accounted for 26%.

The U. S. government was a distant third in both crises. When causes were discussed, 18% of the discussion in the first crisis and 19% in the second pointed to the U.S. government as a cause. The remainder of this discussion centered on the consumer[45] as a cause (5%) and miscellaneous other causes (5%) such as unspecified hoaxes and finite supplies of oil.

OPEC[46]

OPEC was unique among the causes discussed—for while

---

[45] In accordance with a convention generally adhered to in television coverage, "consumer" here refers only to private individuals and excludes industries and corporate entities.

[46] Iran was always aggregated with OPEC, despite the fact that the Iranian revolution, which significantly reduced Iranian oil production, was not a function of OPEC.

Chart 6

Causes of Oil Crises Discussed
[*100% = 1013 Subjects*]

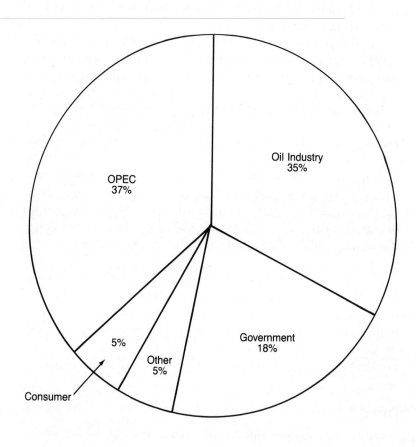

OPEC
37%

Oil Industry
35%

5%

Consumer

Other
5%

Government
18%

the oil industry and the government struggled to deflect blame, OPEC welcomed it. What's more, television rarely questioned OPEC assertions. For instance, ABC reported that, "The Arab states say the reductions will continue until the U.S. follows a more even-handed policy in the Middle East."[47] Coverage of the "oil-weapon" consistently accepted OPEC's claim that embargoes and the like were a political rather than economic weapon.[48] In under 3% of the discussion of the embargo was it noted that occasional shortages in some petroleum product supplies had occurred prior to the embargo.

As the networks eschewed the economic motivation for OPEC moves, their interpretation of OPEC was at times simplistic and contradictory. NBC explained that OPEC "voted to freeze oil prices at their present levels for the next three months as a *good-will gesture*".[49] In the second crisis, the non-economic interpretation again prevailed. A "major unanswered question" reported by CBS' Nelson Benton was, "[H]ow long Saudi Arabia and other producers will pump more oil than normal *to help fill the gap*...."[50] If OPEC production levels were set with helpfulness in mind, their pricing decisions were similarly congenial. Both ABC and CBS carried reports one evening[51] about divisions among OPEC members on the subject of prices. But the reasons "moderates" might wish to set a lower price for their crude than did the "price hawks" (ABC) or "hardliners" (CBS) went unexplained.

Because the networks chose not to portray OPEC members as rational economic actors with explainable economic motivations, they tended to depict them as greedy or capricious. After reporting three new price hikes from OPEC members, NBC's David Brinkley added, "Since all of this bears no relation to the cost of production, or even to the

---

[47] November 5, 1973.
[48] The exact opposite interpretation has, however, been well-argued by Douglas Feith in *Policy Review*, Winder 1981, as well as in *The New Republic*, November 22, 1980.
[49] March 17, 1974 (emphasis added).
[50] January 18, 1979 (emphasis added).
[51] December 15, 1978.

needs of the producers, the only other reason must be greed and gouging."[52] In a similar vein, on CBS John O'Leary, then Deputy Secretary of Energy, lamented, "We find ourselves having simply no say in *an arbitrary pricing action* on the part of OPEC...."[53]

Hoaxes and Price Gouging

In contrast with OPEC's explicit efforts to take credit for the crises, both the oil industry and the government actively tried to deflect blame from themselves. But the oil industry was not nearly as successful as the government in such efforts.

The networks' inability to portray the economic motivations of OPEC was echoed in their inability to convey an economic understanding of the oil industry, and was in part due to the emphasis they placed on the consumption end of the problem. The networks gave scant attention to the issues on which many economists and experts believe the crises hinged. Instead, television entertained at great length a popular notion that the crises were the fault of, and perhaps a hoax perpetrated by, the oil industry.

The possibility that the cause of the crises was a conspiracy among the oil companies, or a hoax of their creation, accounted for 6% of all causes discussed. [See Chart 7] For example, Walter Cronkite reported that "many Americans question the truth of the oil shortage, and the questioning has been apparent in a crescendo of criticism of the big oil companies."[54] The questioning continued in the second crisis with John Chancellor noting that "a lot of people still wonder if the shortages are for real, or part of a scheme to make more money for the oil companies."[55] Senate investigators, according to another NBC report, "accused the American oil companies of acting in concert to increase both prices and profits while the Arab oil producers were raising the costs of their oil."[56]

---

[52] March 6, 1979.
[53] December 20, 1978 (emphasis added).
[54] January 11, 1974.
[55] March 19, 1979.
[56] March 27, 1974.

Chart 7

Discussion of Oil Industry as Cause of Crises

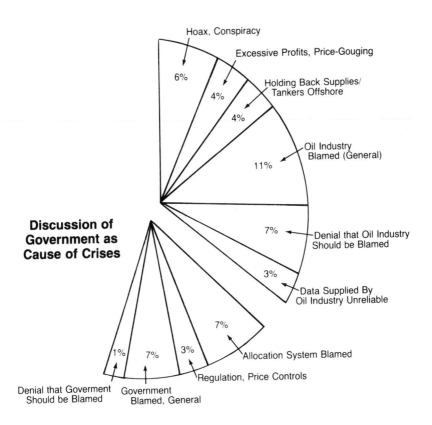

Hoax, Conspiracy

Excessive Profits, Price-Gouging

Holding Back Supplies/ Tankers Offshore

6%

4%

4%

Oil Industry Blamed (General)

11%

Discussion of Government as Cause of Crises

7% — Denial that Oil Industry Should be Blamed

3%

Data Supplied By Oil Industry Unreliable

7%

Allocation System Blamed

1%   7%   3%

Regulation, Price Controls

Denial that Goverment Should be Blamed   Government Blamed, General

A related notion, which constituted another 4% of causes discussed, was that the oil companies were withholding supplies. For example, ABC noted, "There have been reports that despite oil shortages fully loaded tankers are sitting off the Atlantic coast waiting for higher prices before they unload their cargoes."[57] According to an ABC report in the second crisis, "one very simple reason why the lines are long at service stations" might be found in congressional testimony that "the oil companies stockpiled crude oil during the first quarter of this year and refined less gasoline than they could have."[58] Yet another ABC report said that President Carter "asked the Energy Department to find out if the oil companies were causing the shortages by holding out supplies."[59]

Four percent (4%) of all causes discussed dealt with the possibility that the oil industry was making excessive profits through price gouging.[60] For instance, NBC reported, "Senators contended that the companies, given their record profits, find it better to keep prices high rather than negotiate with Arabs for cheaper oil."[61] On ABC, Senator John Durkin was shown exclaiming "How in God's name can you ask the American public to make a sacrifice when you have the horrendous, unconscionable oil company profits for the last quarter?"[62] NBC's Mike Jensen pondered a similar question: ". . . as long as there is evidence that not everyone is sharing in those sacrifices, that some are making a profit while others suffer, then the business of solving the energy problem will be that much more difficult."[63]

Often, the oil industry was blamed in more general terms. Government sources were volubly critical of the oil industry. Congressional hearings were one of the most fertile sources of dramatic film footage. A number of congressmen were shown chewing out oil company executives for the benefit of

---

[57] December 31, 1973.
[58] June 11, 1979.
[59] August 14, 1979.
[60] When legal action was taken by an entity such as the Justice Department, this was noted among charges of illegality in "other subjects."
[61] March 27, 1974.
[62] February 28, 1979.
[63] March 19, 1979.

the folks back home. For example, in scolding an Exxon executive, Senator Henry Jackson said on CBS: "These are childish responses Mr. Baze ... I'm, I'm flabbergasted."[64] In the second crisis, Senator Jackson explained why he doubted the windfall profits tax would pass: "I've been fighting with the oil lobby, and I'm telling you, the word is greed."[65]

On CBS, Walter Cronkite suggested that "the root of the problem" was in "the very nature of the giant oil companies."[66] But not every story pointed the finger so explicitly. Some reports simply noted the correlative rise in fuel prices and oil company profits. One evening CBS noted this connection (*i.e.*, "...oil profits which are soaring like the prices at the gas pumps"[67]) no less than three times in the same story.

Discussion of the oil industry as a cause did not always exonerate everyone else. Reports in both crises discussed the possible effects of oil company campaign contributions both to President Nixon[68] and to members of Congress.[69] On CBS Ralph Nader offered his views: "I think it is an energy crisis for consumers who are being subjected to billions of dollars of unarmed robbery by the oil companies in collusion with government support."[70] A good conspiracy theory could explain nearly everything, as did this one by Christopher Rand whom CBS characterized as a mideast oil expert and former oil company employee:

> ... the [oil] companies are withholding their supplies from us with the blessings of the Federal government ... certainly the companies want to raise prices, and certainly the Administration wants to create an artificial crisis to divert public attention from Watergate and perhaps also to ... lay the groundwork, for some sort of recessionary

---

[64] January 23, 1974.
[65] ABC, April 6, 1979.
[66] November 29, 1973.
[67] April 23, 1979.
[68] NBC, December 13, 1973.
[69] CBS, April 23, 1979.
[70] January 14, 1974.

measures that the government would not like to be directly responsible for taking.[71]

One-fifth of the discussion of the oil industry as a cause consisted of denials that the industry should recieve such blame. For example, Z. D. Bonner, president of Gulf Oil Corp., said on CBS:

There are, I think, three mistaken notions about this energy shortage. First is the notion that the Arabs are to blame for the U. S. energy crisis. Second is the mistaken notion that U. S. energy resources are about to be exhausted. And third, the notion that the major oil companies contrive to produce shortages so they can get greater profits.[72]

Nor did all such denials come from the oil industry. ABC, for example, reported that the Energy Department had released a 50-page report clearing the [oil] industry of hoarding supplies to hold out for higher prices."[73]

That particular story proceeded to quote a number of criticisms of the report as "misleading" and "an exercise in propaganda" before concluding that: "Press Secretary Jody Powell ... indicated that any discrepancies would be changed in the final version. This could well be a graceful way to turn around a report that may be based only on information provided by the oil companies."[74]

The credibility of the oil companies was called into question in other instances, as in this CBS report: "Oil industry figures are not to be trusted. That's what a panel of nongovernment experts told the subcommittee."[75] Controversy surrounding the reliability of oil industry data accounted for 3% of all causes discussed. Questions concerning the credibility of the oil industry had wider implications, of course. If the industry could not be trusted to assess the crises accurately, then its recommendations for ending the crises were at least as suspect.

---

[71] January 11, 1974.
[72] January 21, 1974.
[73] August 14, 1979.
[74] *Ibid.*
[75] January 16, 1974.

Government as a Cause

In both crises, the government was the subject of less than 20% of all causes discussed. Blame came in two different forms: the government was discussed as a cause because it did not regulate enough and because it regulated too much. For example, CBS reported that both the Air Resources Commissioner of New York City and Consolidated Edison "sharply criticized the Nixon Administration for failing to provide nationwide allocation of oil supplies."[76] The government's allocation system was the subject of 7% of all discussion of causes. [See Chart 7]

The government's failure "to police the [oil] industry,"[77] and "the government's failure to do something about [New England's home heating oil supply shortages]"[78] were likewise areas in which the government was accused of "inaction."[79]

On the other hand, overly ambitious government regulators were discussed as a cause. For example, ABC reported that Shell Oil had reduced deliveries and "cutbacks are blamed mainly on Washington."[80] A Shell spokesman appearing in the story further cited "regulatory constraints limiting the company's alternatives."[81] NBC's Robert Goralski noted that the chairman of the Price Commission was responsible "for controlling petroleum prices, keeping them down which discouraged energy expansion."[82] Walter Cronkite observed: "The oil companies say government policies hinder the building of refineries. They cite environmental restraints."[83] And Roger Peterson reported on abuse of government regulations concerning "new" versus "old" oil, and concluded, "What all this shows is the regulations covering oil production for the past five years haven't worked and that's cost us billions."[84]

---

[76] October 26, 1973.
[77] CBS, November 29, 1973.
[78] NBC, December 13, 1973.
[79] *Ibid.*
[80] December 1, 1978.
[81] *Ibid.*
[82] December 13, 1973.
[83] November 29, 1973.
[84] November 23, 1978.

Government as a cause was also portrayed as an inter-
necine battle. For example, CBS' Phil Jones, referring to
Republican minority leader Howard Baker, said that "if the
voters ask [Baker] who's responsible for no energy legislation he will blame the Democrats."[85] And on NBC President Nixon declared "... now it's time for the Congress to
get away from some of these other diversions, if they have
time, and get on with the energy crisis."[86]

The consumer was also discussed as a cause. In the first
crisis, such discussion amounted to only 2% of all causes
discussed, and centered on "wasteful consumption of
fuel."[87] On December 20, 1973, all three networks led off
their broadcasts with coverage of "energy czar" William
Simon's announcement of a voluntary ten-gallon maximum
on gasoline purchases.

In the second crisis, consumers were discussed as a cause
much more frequently, accounting for 9% of all causes discussed. Consumer profligacy was still at issue (*e.g.,* "If
Americans would just stop wasting energy"[88]), but the focus
had shifted to the practice of "topping-off" gas tanks. CBS
broadcast one description of this phenomenon from Senator Abraham Ribicoff: "... the public is so scared ... so
frightened about tomorrow that every time they pass a gasoline station they fill up the gas; they're just like a dog beside
every telegraph pole."[89] On May 15, 1979, both ABC and
CBS led off their broadcasts with stories on an emergency
anti-tank-topping regulation announced by the Energy Department which would allow dealers to set minimum pur-
chase requirements. One question which escaped mention
in the networks' coverage of waste and hoarding was why
the price of oil and its value were not in equilibrium to begin
with.

Finally, 5% of the discussion of causes fit into none of the
above categories. The limited supply of oil both worldwide
and in this country was blamed quite a number of times. For

---

[85] August 2, 1979.
[86] December 13, 1973.
[87] *Ibid.*
[88] Roger Peterson, ABC, February 27, 1979.
[89] January 21, 1974.

Chart 8

Sources* of Discussion of
Causes of Crises

[*100% = 878 Identifiable Sources—see footnote*]

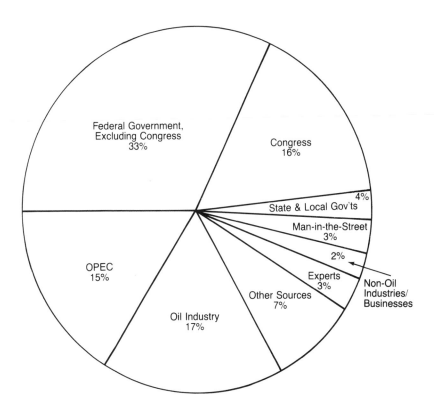

*unattributed sources excluded (135 causes, or 13% of all causes discussed,
were provided by unidentifiable sources)

instance, on CBS, Energy Secretary Schlesinger stated that "the notion that there is a plentiful supply of oil is a mirage."[90] Non-oil industries were also accused of stockpiling oil and aggravating the shortage. Additionally, some talk of a hoax never specified the perpetrator. NBC reported that George Meany said, "The oil shortage may have been at least partly contrived,"[91] with no further elaboration. Martin Lobell, a former Senate energy aide, was shown on CBS testifying on Capitol Hill that "the shortage has been manipulated."[92] Walter Cronkite reported on a man who "put some of the blame [for the nation's energy crisis] on the women's movement. [He] said that as more women enter the workforce they increase energy use in factories and offices. And furthermore, they make more money so their families can spend more on such things as electrical appliances."[93] Comment from the National Organization for Women was included in the report.

As Chart 8 shows, the government was the major source of information when causes were discussed—three times more often than either OPEC or the oil industry. This should be compared with Chart 6 which shows that the government's role (if any) in causing the crises was discussed just 18% of the time.

---

[90] August 1, 1979.
[91] March 13, 1974.
[92] January 16, 1974.
[93] January 29, 1974.

IV. Effects

Television's presentation of the effects of the oil crises was also given from the "consumer's" standpoint. This approach, however, seemed much better suited to a delineation of effects (*i.e.*, how the crises affected the ordinary American) than it was elsewhere.

Coverage of effects presented information from a spectrum of sources much broader and more diverse than the narrow universe of sources that appeared in coverage of solutions and causes. This occurred largely because the government did not dominate the coverage of effects. Whereas government sources accounted for 77% of all solutions discussed, and 53% of all causes, the government was the source of just 30% of the discussion of effects. [See Chart 9] The networks also sought out non-oil industry and labor sources and the man-in-the-street. For some of these sources, discussion of the oil crises was largely confined to this subject of effects. Seventy-three percent (73%) of the information from non-oil industries dealt with effects, and 56% of what the man-in-the-street discussed concerned effects. Service station operators, who discussed 24% of all subjects raised by the oil industry, accounted for 84% of what the oil industry said regarding the effects of the crises. Over half of what service station operators said in total was on the subject of effects.

Another distinctive characteristic of the coverage of ef-

Chart 9

Sources* of Discussion of Effects
[100% = 869 Identifiable Sources—see footnote]

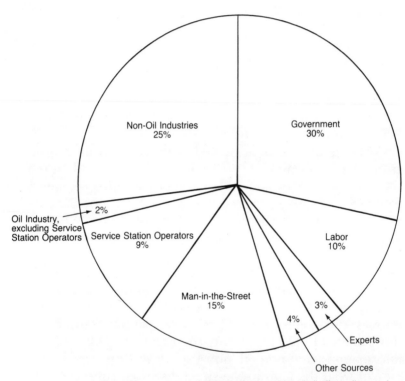

*unattributed sources excluded (413 effects, or 32% of all effects discussed, were provided by unidentifiable sources)

fects was that a high percentage of the subjects 32%, was not attributed to any source. The percentage of effects with unidentifiable sources went from 27% in the first crisis to 44% in the second. In fact, when a reporter did not attribute his information, it was often in speaking of effects. Forty percent (40%) of all topics with unidentifiable sources were concerned with effects of the crises. [Refer back to Table 3]

To convey the dimensions and ramifications of the crises, television showed a variety of different matters. Coverage of the adverse impact of the crises on industry constituted 29% of the effects discussed. [See Chart 10] Adversely affected industries received twice as much coverage in the first crisis as in the second (35% vs. 17%). The conditions at service stations, on the other hand, was a much bigger story in the second crisis, increasing from 20% to 38% of all effects discussed. The public's reaction to the crises also received substantial coverage (16%), as did effects reported in broad macroeconomic terms, such as effects on inflation, unemployment and the like. In addition, the networks covered a diverse assortment of effects that fit in none of these categories.

The effects of the oil crises on a variety of businesses and industries received substantial coverage [See Chart 11] The trucking industry, for instance, was doubly distressed, both by higher diesel fuel costs and by lowered highway speed limits, and truckers' protests were featured on television in both crises. When the truckers blockaded major interstate highways, television crews took to the air for footage of the chaotic array of tractor-trailers and the resultant snarl of traffic. When state troopers attempted to clear the highways, the confrontations with truckers also made the news. And when truckers shot at each other, such incidents were likewise well-covered.

Both the airline and auto industries were also covered extensively due to the crises. Typical was a report by NBC's Norma Quarles which began: "At O'Hare airport and other airports around the country, TWA, National, United and Delta airlines have cancelled flights because they are short of jet fuels."[94] Coverage of the auto industry frequently noted

---

[94] March 16, 1979.

Chart 10

Effects of Crises Discussed
[*100% = 1282 subjects*]

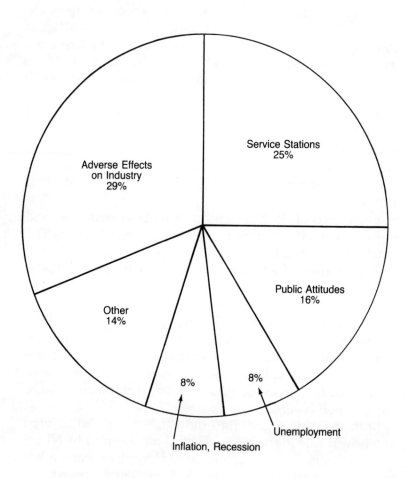

Service Stations
25%

Adverse Effects
on Industry
29%

Public Attitudes
16%

Other
14%

8%

8%

Inflation, Recession

Unemployment

Chart 11

Adverse Effects on Industries

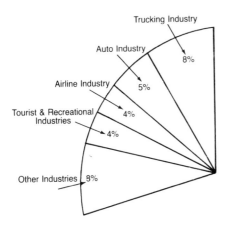

Trucking Industry 8%

Auto Industry 5%

Airline Industry 4%

Tourist & Recreational Industries 4%

Other Industries 8%

Service Stations

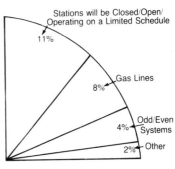

Stations will be Closed/Open/Operating on a Limited Schedule 11%

Gas Lines 8%

Odd/Even Systems 4%

Other 2%

Public Attitudes

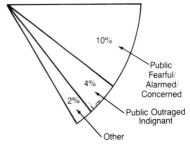

10% Public Fearful/Alarmed/Concerned

4% Public Outraged Indignant

2% Other

a decline in the sales of big cars and a surge in the sale of small ones.

In both crises, the impact of the oil crises on the tourism and recreation industries was covered extensively. One report by NBC's Don Oliver concluded in this way: "We wanted to show that Disneyland and seaside and mountain resorts have also had a loss of business. But you'll have to take our word for it. We didn't have enough gas to get to those places."[95]

The adverse effects of the crises on both utilities and service stations were also reported extensively. Mentioned in this context as well was a wide assortment of adversely affected industries—textile, recorded music, toy, plastic and slaughterhouse industries, to name but a few.

The effects of the crises on service stations commanded fully 25% of all coverage of the effects. The question of whether stations would be open or not accounted for 11% of the effects discussed. NBC's Jessica Savitch led off one broadcast with this: "In many parts of the country today, Mother's Day visits depended on whether mother lived less than a full tank of gas away. In most states up to half of the service stations were closed."[96]

Aerial footage of interminable gas lines wrapping around city blocks was another favored story. People waiting in gas lines were often interviewed for the evening news. Typical of their comments was the observation of one irked motorist broadcast on CBS: "Obviously this is ridiculous. It's wasting everyone's time."[97]

The attitude of the public toward the crises also received considerable attention. In the first crisis, the public was generally depicted as fearful, concerned or alarmed. On CBS, Representative Silvio Conte stated that his constituents "come in there with trembling hands and show me their gas bill and their oil bill and say, 'Mr. Conte, I—I just can't exist....'"[98] According to Senator Charles Percy, "The

---

[95] May 13, 1979.
[96] May 13, 1979.
[97] June 19, 1979.
[98] January 16, 1974.

American public doesn't know who or what to believe."[99]

In the second crisis, the networks portrayed a more indignant and outraged public. As an unidentified New Yorker put it, "I think it stinks."[100]

About 16% of the coverage of effects in each crisis centered on a variety of macroeconomic measures. Unemployment resulting from the crises was given substantial coverage, particularly in the first crisis. For instance, on NBC:

> The government said today that the fuel shortage is beginning to cause unemployment. The Labor Department said that almost 20,000 people who lost their jobs because of the energy crisis are now applying for or getting unemployment insurance.[101]

Frequently, unemployment was discussed with respect to specific industries or firms. For example, both CBS and NBC reported that, due to the fuel shortage, Disney World was laying off 2,000 employees. Among them, John Chancellor noted, were understudies for Mickey Mouse and Donald Duck.[102]

The effect of the crises on inflation and its possible recessionary impact was also reported, particularly in the second crisis. Often a variety of economic indicators was reported. CBS one evening mentioned increases in the cost of living, fuel costs and unemployment among the "grave economic effects expected to result from the crisis."[103]

Finally, extensive coverage was given to an assortment of other effects. All three networks reported numerous times what Walter Cronkite termed "one bit of silver lining in the whole mess."[104] The lower highway speed limit had substantially reduced the death toll. Other positive effects were reported, such as the increased business for chimney sweeps and for manufacturers of wood stoves, woolen garments,

---

[99] CBS, January 21, 1974.
[100] CBS, June 19, 1979.
[101] December 20, 1973.
[102] NBC, January 10, 1974.
[103] December 11, 1973.
[104] January 14, 1974.

and gas pumps (the old pumps couldn't register $1.00/gallon for gasoline, so new ones had to be manufactured).

On the other hand, those particularly hurt by the crises were also discussed. All three networks reported one evening on Senate subcommittee hearings concerning the effects of the oil crises on the poor and elderly.[105] Declining revenues from state fuel taxes were also noted a number of times.

A number of stories dealt with the possible legal and health ramifications of the crises. NBC reported that the Trial Lawyers Association warned: "those motorists who carry gasoline cans in their cars could be held liable for accidents that result in fires or explosions...."[106] Another story focused on the problems, both criminal and medical, surrounding the practice of siphoning gas. One young woman who had ingested some gasoline in the process explained, "I just swallowed too much...."[107]

The networks' portrayal of the effects of the crises was clearly oriented toward stories that were entertaining and visually appealing.

---

[105] January 22, 1974.
[106] January 8, 1974.
[107] NBC, July 5, 1979.

V. Other Subjects

Not all coverage of the oil crises centered on solutions, causes or effects of the crises. Often, changes in prices or in the levels of supply, production or demand were reported without reference to their possible role in causing or solving the crises. Other matters, such as oil company profits and illegal activity on the part of the oil industry, were similarly reported. [See Chart 12]

Information on these "other subjects" came predominantly from the government, the oil industry and OPEC. [See Chart 13] In fact, 48% of what OPEC discussed was of this sort, as was 48% of what the oil industry (including service station operators) discussed. Fifty-nine percent (59%) of everything discussed by the oil industry (excluding service station operators) was on these other subjects. So while the networks relied heavily on the government for information on solutions, the networks relied on the oil industry for information on matters apart from solutions, causes or effects.

One-third of the discussion of these other subjects dealt with the actual status of the shortage, or with the level of oil supplies in the country. [See Chart 14] For example, CBS' Nelson Benton reported one evening that "energy chief William Simon assured the [joint congressional] subcommittee that the crisis is real; there is an oil shortage...."[108]

---

[108]January 14, 1974.

Chart 12

Other Subjects* Discussed
When not discussed as a solution, cause or effect of the oil crises
[100% = 1445 Subjects]

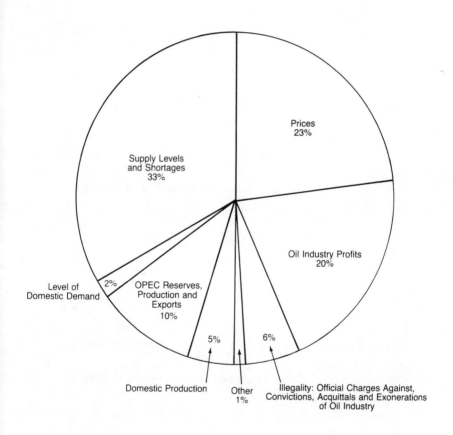

Prices
23%

Supply Levels
and Shortages
33%

Oil Industry Profits
20%

Level of
Domestic Demand

2%

OPEC Reserves,
Production and
Exports
10%

5%

6%

Domestic Production

Other
1%

Illegality: Official Charges Against,
Convictions, Acquittals and Exonerations
of Oil Industry

Chart 13

Sources* for Discussion of Other Subjects
[100% = 1189 Identifiable Sources—see footnote]

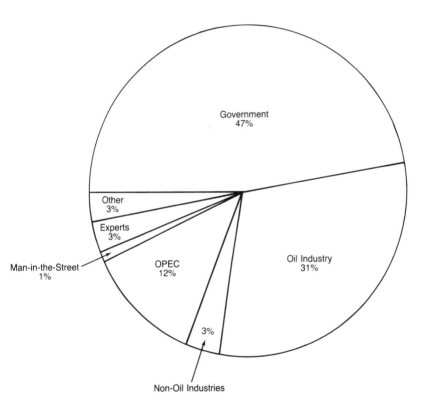

*unatrributed sources excluded (256 other subjects, or 18% of all
other subjects discussed, were provided by unidentifiable sources)

Similarly, the president of the American Petroleum Institute was shown on CBS stating, "It's a very real and accute shortage, and its one that'll be with us after the embargo's lifted...."[109]

It was also reported with some regularity that the shortage was not as bad as expected, or that it was expected to end. Sometimes, contradictory statements about the shortage appeared in the same broadcast. Walter Cronkite began one broadcast: "It's hard for the public to know what to think." He went on to explain:

> In spite of reductions in gasoline deliveries to stations and industry warnings that the shortage is real and continuing, an Administration official now comes forward to say he thinks the summer won't be too bad, only slightly less gasoline than last summer."[110]

Specific levels of oil supplies were also reported. For example:

> The American Petroleum Institute said that, compared to the previous week, oil imports dropped 12%, and gasoline stocks dropped some 3.6 million barrels, or about 2%. Crude stocks were also down...."[111]

Increases in the level of supply were likewise given coverage, as in this report from NBC's Robert Hagar: "The American Petroleum Institute said late this afternoon that gasoline stocks in the U.S. have increased some for the second week in a row."[112]

A second important subject that was frequently discussed without reference to its possible role as a solution to, cause or effect of the crises was the price of oil, which was generally on the rise. Price rises were usually reported without any particular context, somewhat the way the Dow Jones closing averages are reported each night. For example:

> ... the November jump in wholesale oil prices was dramatic and disheartening—almost 35%. A breakdown

---

[109] CBS, January 11, 1974.
[110] June 1, 1979.
[111] CBS, January 16, 1974.
[112] December 6, 1978.

Chart 14

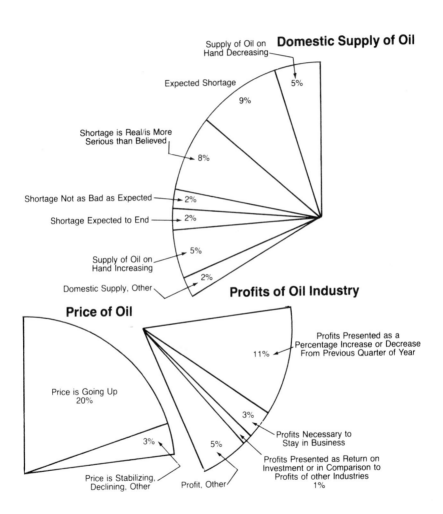

Domestic Supply of Oil

Supply of Oil on Hand Decreasing — 5%

Expected Shortage 9%

Shortage is Real/is More Serious than Believed 8%

Shortage Not as Bad as Expected — 2%

Shortage Expected to End — 2%

Supply of Oil on Hand Increasing 5%

Domestic Supply, Other 2%

Profits of Oil Industry

Profits Presented as a Percentage Increase or Decrease From Previous Quarter of Year 11%

Profits Necessary to Stay in Business 3%

Profits Presented as Return on Investment or in Comparison to Profits of other Industries 1%

Profit, Other 5%

Price of Oil

Price is Going Up 20%

Price is Stabilizing, Declining, Other 3%

showed gasoline prices up 32%, diesel and home heating fuel up 45% last month."[113]

At times the discussion of prices without reference to causes or solutions appeared lacking. For instance, John Chancellor reported that the Cost of Living Council "*approved* another increase in retail gasoline prices to reflect higher wholesale costs." The Council "also *approved* increases in diesel and home heating fuel costs...."[114] In another broadcast, Chancellor reported that the Council "announced today that the price of domestic crude oil *will be allowed to rise* 23%, which means that higher prices are on the way...."[115] Missing from these reports was any explanation of why a government agency was "approving" or "allowing" price rises.

Levels of oil production, both in the United States and in OPEC countries were at times discussed without reference to causes, effects or solutions. For instance, NBC reported that "... officials in Iran said they expect exports to resume on Monday with production rising to about half the pre-Khomeini level within a month."[116]

Again, the matter-of-fact discussion of such subjects without reference to causes or solutions was at times puzzling. ABC's Roger Peterson reported that: "There were no gas lines in 1960 ... the United States produced more oil than it could use. Energy supplies seemed inexhaustible. But times have changed."[117] Peterson went on to describe the origins of OPEC, then stated: "But *the oil surplus slowly evaporated*. By 1973, *no longer self-sufficient*, we were importing 35% of our oil...."[118] At no point in this nearly three-minute synopsis of the oil crises were price controls mentioned. Physical rather than economic laws provide the explanation for changes in the level of domestic oil production—the oil just evaporated!

Discussion of the supply levels, prices and production

---

[113] NBC, December 6, 1973.
[114] October 15, 1973 (emphases added).
[115] December 19, 1973 (emphasis added).
[116] February 27, 1979.
[117] June 26, 1979.
[118] *Ibid.* (emphases added).

levels of oil, when not treated as solutions, causes or effects, was generally provided to measure the dimensions of the crises themselves. In addition to such subjects, coverage was also given to two subjects directly related to the oil industry: first, the profits of the oil industry, and in particular of the major oil companies; and second, charges of illegality against some element of the oil industry.

More often than not, profits were reported as a rise over the preceding quarter or year, as in this CBS report:

> ... Three more major oil companies have released their 1973 profit statements, and, like Exxon yesterday, they reported good gains over '72. Mobil profits were up 47% after taxes; Texaco profits, up 45%; and Shell up 28%. And for the fourth quarter alone, after the start of the Arab oil embargo, Mobil and Texaco both reported gains of almost 70%.[119]

In the second crisis, the same style of coverage was predominant, with Walter Cronkite stating: "Another big firm, Mobil, has joined the parade of oil companies reporting large profits for the first three months of this year. Mobil reported profits of $473 million, up 81% from the same period last year."[120]

Profits were also discussed as being necessary for the oil industry to stay in business, as in this exchange between an unidentified CBS interviewer and Exxon Board Chairman J. K. Jamieson:

> INTERVIEWER: We all seem to feel that oil is an essential resource, it's essential to our economy and the well-being of the people. Doesn't the oil industry have a special responsibility?
>
> JAMIESON: For what?
>
> INTERVIEWER: For providing supplies regardless of cost in given emergency situations.
>
> JAMIESON: Within somewhat narrow limits I would say yes, but we certainly couldn't

---

[119] January 24, 1974.
[120] April 26, 1979.

continue to do that for any length of time or *we would go broke.*[121]

Five percent (5%) of the discussion of profits was in terms of return on investment or in comparison to the profitability of other industries. For example, William Simon was shown on CBS asserting that oil industry profits "are not exorbitant in looking at the rest of the free-enterprise system in this country."[122] Nelson Benton elaborated on this position in the second crisis saying: "[the oil companies] argue they make about 14 cents a year on each dollar of their capital, compared with about 15 cents earned by the rest of U.S. industry."[123]

The remaining discussion of profits fit in none of these categories. For example, Deputy Secretary of Energy John O'Leary was shown on NBC making reference to "the charge and reality of obscene profits."[124] While not a favorable portrayal of oil industry profits, such comments did not explicitly discuss profits as a cause of the crises.

The second subject which directly involved the oil industry was concerned with legal proceedings against various elements within the industry. Such charges ranged from: "The Energy Department has accused twenty Florida service stations of price gouging on gasoline"[125] to "The fifteen major oil companies overcharged their customers by at least $4.8 billion from the Arab oil embargo until 1976"[126] according to Energy Department investigators. Also reported, with less frequency, was the resolution of such charges, in conviction, acquittal or whatever.

Finally, some of the coverage of the oil crises defies classification in any of the categories so far discussed. On NBC John Chancellor addressed a thorny problem concerning rationing one night without reference to its role as a solution to the oil crises. Instead, he discussed the two different ways

---

[121] November 29, 1973 (emphasis added).
[122] January 23, 1974.
[123] CBS, April 23, 1979.
[124] March 19, 1979.
[125] CBS, April 20, 1979.
[126] ABC, August 14, 1979.

the word "rationing" was frequently pronounced (*i.e.,* to rhyme with stationing or with fashioning).[127]

*   *   *

The subjects that were not portrayed as causes, solutions or effects are notable, for in many cases they dealt with the economic components of the crises—prices, supply levels and production levels. It is revealing that the networks treated the economic components of the crises as incidental to the solutions and causes, rather than as the very essence of both.

---

[127] December 20, 1973.

Appendix—Methodology

This study would have been impossible were it not for the Vanderbilt Television News Archive. Since 1968, Vanderbilt has been videotaping the evening news broadcasts of ABC, CBS and NBC. In addition, it publishes monthly abstracts and an index through which researchers can identify and place orders for stories of interest. (Since 1979 The Media Institute has used the Vanderbilt Archive as the source for data on business and economic stories, which the Institute publishes in its fortnightly monitor, *The Television Business-Economic News Index.*) Vanderbilt then compiles videotapes of those stories and loans out the tapes for a modest fee. Thanks to the excellent services the Vanderbilt Television News Archive provides, The Media Institute was able to identify and obtain videotapes of every network evening news broadcast relevant to this study, a total of 39½ hours of germane tape.

Criteria for Selection of Stories

The Vanderbilt Index was used to identify oil crisis stories. All stories appearing under "Energy Crisis" were included, with the following exceptions:

1. *Alternate Fuels.* Only stories pertaining to crude oil and refined products, *e.g.,* gasoline, heating oil, aviation fuel, and kerosene were included. Thus, stories concerning oil shales, synfuels, natural gas, coal, solar, geothermal, and

nuclear energy were excluded.

2. *Foreign Events Only Indirectly Related to the Oil Crises in the U.S.* Thus, stories about the Arab-Israeli War, the Iranian Revolution, and the energy crises in foreign countries were excluded, unless they referred to the production of petroleum for export to the U.S. If a story mentioned oil production in the context of a longer story about other matters, as in a story about the Iranian Revolution that makes mention of cutbacks in production, then only that portion of the story making a direct reference to oil was included.

3. *Bureaucratic Shuffles.* Stories noting, for example, that the Federal Energy Administration had had a change of command were excluded, unless some substantive policy matters were discussed in the story.

4. Stories about collisions, accidents, safety, or pollution were excluded.

5. *Editorials,* such as Howard K. Smith's "Commentaries" on ABC, Eric Sevareid's "Analysis" on CBS, and David Brinkley's "Journal" on NBC, were excluded.

Researchers determined that the coverage of oil crisis stories began in earnest in October 1973 (the OPEC embargo was declared October 17), and tapered off in May 1974. The oil crises again became a sizeable story on the evening news in November 1978, and did not wane until August 1979.

The Coding Process

The 39½ hours of videotape were analyzed by The Media Institute researchers employing a quantitative and objective technique known as content analysis. This is a systematic and reliable method for organizing communication content into various categories. The coding sheet contained pre-established categories into which all information was coded. The coder answered a variety of questions by viewing the film several times and applying the decision rules to the particulars of each story. Checks on the reliability of the coders were conducted, and exceeded acceptable minimums.

A separate code sheet was devoted to each story. Because several related reports often follow consecutively in a newscast, it was sometimes difficult to determine when one story had ended and the next begun. The coder therefore referred to the Vanderbilt Abstracts, wherein the beginning of each new story is indicated by a notation of the time at which it was originally broadcast.

A number of questions were designed to establish the identity of the subject (in this case, the individual story). Vanderbilt greatly facilitates this effort by electronically coding its videotapes so that the network, date, and time of broadcast (to the nearest ten seconds) is displayed across the top of the screen.

Within any one story, matters not directly related to the oil crises were at times discussed. To avoid muddying the waters, researchers deleted from the total time given each story any tangential material that continued for thirty seconds or longer without mention of matters germane to the study.

A preliminary survey of the videotapes, employing a modified form of emergent analysis, had resulted in the identification of all principal segments of society to appear in news coverage of the crises. For those segments of society, such as the government, which appeared or were quoted frequently in the news, further subcategories, such as Congress or state and local government, were established. This same process also resulted in the identification of the major categories and subcategories of coverage. As a result, the researchers had a list of 60 categories that accommodated all sources of information, and a second list of 200 categories that accommodated all subjects discussed.

In viewing each story, coders noted each subject that was raised, and the source of such subjects. An individual story might be adequately summarized by one such source-subject combination, or might require two dozen or more such combinations, depending on how complicated a story was and how much information it contained.

The data collected in this fashion was then analyzed with the assistance of an IBM 3033 at the George Washington University Center for Computing in Washington, D.C. In this way the data was aggregated and disaggregated in a wide variety of ways.

DEMCO